The Vailan or ann

A synopsis of Prof. I. N. Vail's arg
the claim that this Earth once possessed a Saturn-like
system of rings

Stephen Bowers

(Contributor: Isaac N. Vail)

Alpha Editions

This edition published in 2024

ISBN : 9789362099518

Design and Setting By
Alpha Editions
www.alphaedis.com
Email - info@alphaedis.com

Contents

PREFACE.

The theory advanced by Prof. I. N. Vail accounts for the formation of the earth's crust, with its associated minerals, in the fact that it was once surrounded by rings of aqueous vapor, containing much of its present solid matter, which fell as mighty deluges. The last of these rings descended at the time of the Noachian deluge and caused that catastrophe, which is so graphically described by Moses, and which tradition has sung in the ears of every tribe of Adam's race. The formation of these rings was caused by the intense heat, which drove to an immense distance every substance which could be reduced to vapor, and where they formed as annular bands or rings similar to those surrounding the planet Saturn at the present time. After long ages the portion nearest the earth slowly overcanopied the heavens, and owing to the lack of centrifugal force began its descent at the poles.

This theory explains certain phenomena better than any other yet advanced by scientists. It accounts for the uplift of mountains; the deposit of coal and other minerals; the glacial age; the retardation of the moon, and it alone explains much contained in the first eight chapters of Genesis.

Prof. Vail has published a volume of about 400 pages on this subject, which for clearness of statement and logical conclusions has seldom been equaled by previous writers on scientific subjects. He deals in convincing facts which are destined to overturn many pre-conceived theories in the science of geology.

My object in sending forth this pamphlet is to call the attention of intelligent readers to a theory which must engage the attention of scientists in the future, and which will enable the geologist to make clear many things which are now obscure. I respectfully ask for the following pages a candid reading, and for further information on the subject refer the reader to Prof. Vail's "Story of the Rocks", and to other works of the gifted author, which are now passing through the press.

S. B.

VENTURA, CALIFORNIA,

September 1, 1892.

THE VAILAN OR ANNULAR SYSTEM.

IMPORTANCE OF THE QUESTION.

Jupiter's belts are doubtless aqueous vapor driven from that planet by heat; similar in every respect, probably, to the primitive condition of our globe. This vapor would not all fall at once on the cooling of the earth, but the upper portion would continue to revolve for a long period.

All geologists agree that the earth was once in an igneous fluid state, and during that condition all of its waters and whatever else could be vaporized and sublimed by heat, as the less refractory metals and minerals, were driven away from its surface. The foundation of the Annular System was the molten or igneous world. The vaporized water, mineral and metallic elements repelled from it existed as a great vaporized atmosphere that rotated with the earth.

If the earth then rotated once in twenty-four hours, so did the atmosphere. Proctor and some others claim that the earth then rotated in three hours; if so, the atmosphere did the same. No matter how long or how short the period of the earth's rotation, the upper vapors rotated with it. Then, when and how did these vapors and other materials composing the atmosphere return to the earth? Geologists generally have claimed that they fell at the close of the igneous period; but the Annular Theory claims that they did not, and it undertakes to explain the phenomena of the geologic ages and epochs upon this claim.

The most eminent scientists agree that the vapors were driven off at least 200,000 miles from the earth, and many claim a distance of 240,000 miles. All of the carbon in the grand casement of aqueous rocks, the vast oceans of oxygen now contained in the silicates, sulphates, carbonates and oxides of the crust, as well as the nitrogen and hydrogen in numerous compounds enormously swelled its volume. But the Annular Theory will claim but 100,000 miles as the atmosphere and that the earth rotated as now, once in twenty-four hours. At the equator it revolves at the rate of 1,000 miles an hour, at which rate the periphery of the earth's primitive atmosphere would revolve more than 25,000 miles an hour.

Now it is mathematically certain that a body in our atmosphere revolving at the rate of 17,500 miles an hour could not fall to the earth's surface. By Kepler's "Third Law" we can readily demonstrate not only that these vapors were thrown out into a ring system, but how far beyond the earth they reached, namely: "The squares of the periodic times of revolving satellites are proportioned to the cubes of their mean distances from the primary around which they move."

The vapors nearest the earth did not possess the energy of satellites, consequently they fell to the earth, as the latter's surface cooled, leaving the more distant matter moving independently above it.

EVIDENCES OF THE GEOLOGIC RECORD.

When the earth was in a state of fiery fluidity, all of the water it now contains was suspended at a great distance above it. Beside the oceans which now cover three-fourths of the surface of the globe, rocks and coals contain from ten per cent to one half water, all of which was primarily held in suspension. The bosom of the earth is continually absorbing water as is demonstrated by deep mines and other excavations. Dana estimates that even if the crust of the earth is but five miles thick that the oceans would be 400 feet deeper if all of the earth's imbibed waters could be returned to them. But the earth's crust is more likely to be 100 miles thick, and it has been imbibing these waters for millions of years if not millions of ages. This would increase the oceans to about 8000 feet deeper than now. Yet oceans are much deeper today than they were in geologic times.

This great mass of vapor would rotate by centrifugal force at the equator, but there being no such force at the poles it was there kept from falling by heat alone. If the earth had not rotated the vapors would have occupied great heights; but centrifugal force being aided by actual rotation they were driven much farther. These forces necessarily drove the vapors over the equator. If, however, any vapors were left at the poles they must have fallen when the earth cooled down.

At that age rolled the first born ocean around the globe. Clouds formed, rain descended, and winds swept the earth. There was summer and winter, and day and night.

The centripetal force of the rings was gradually retarded by the influence of the moon, and the gravital force was increased until the rings spread over the earth or approached it. When the innermost ring gradually descended toward the earth and came in contact with the air it was checked, and necessarily spread out toward the poles. Gravital force is strongest in the polar regions. If the rings of Saturn and Jupiter could increase their motion they would rise to greater heights. If they could become slower they would sink toward the poles.

EVIDENCE FROM OTHER PLANETS.

We have never seen the actual face of Saturn, and the sun is never visible to its inhabitants. It is a planet upon which there is probably perpetual day. The belts are composed of the same kind of material as the super-crust of the earth—silicious, calcareous and carbonaceous matter. They will in time become a part of the planet's sedimentary formation.

When the inveterate fires of the sun shall have died out, forms of carbon and associated forms of aqueous and mineral matter will form an annular system around it.

A burning world must be a smoking world, and from its furnaces must arise vast volumes of unconsumed carbon to mingle with suspended vapors.

When Saturn's rings fall to the body of the planet its moons will necessarily retire a little farther from it. Astronomers say that our moon is gradually retiring from the earth. Then it must have had an annular system which fell and caused the moon to recede.

FURTHER EXAMINATION OF THE RECORD.

The vapors contained silex, quartz and whatever else was vaporized and suspended therein. After the atmosphere had cooled it deposited on the earth what it contained when heated. Much of the sedimentary beds built upon the Laurentian and older rocks were simply precipitated from the annular system.

Iron and sulphur existed in the upper ocean as metallic and mineral salts. In the cooling process the heavier minerals and metals would necessarily locate nearest the earth and be the first to fall. True they were disseminated to a certain extent throughout the system.

Iron and other heavy metals formed beds in the sea bottom. Iron from Iron Mountain, Mo., and Pilot Knob, also lead and copper ores are in the Laurentian rocks. These rocks are aqueous or sedimentary. The annular matter fell but in small part in equatorial regions, but largely in temperate and frigid zones.

It is folly to suppose that all the matter of aqueous beds were deposited from previous aqueous beds by denudation. How were subsequent lime deposits made from silicious Archaean beds? Denudation has taken place in all ages, and a fall and precipitation of exotic matter—tellurio-cosmic matter—aided in the work.

CONCLUSIONS REACHED.

1. All terrestial waters were held in suspension.

2. This rotated as a part and parcel of the earth—a primeval atmosphere of great complexity of material.

3. This suspended matter gathered in the earth's equatorial heavens, and on condensing contracted and segregated into rings which revolved independently.

4. The waters on high fell in a succession of stupendous cataclysms.

5. The first ocean was impregnated with mineral and metallic salts.

6. It required a vast lapse of time for rings to fall. Each ring continued to revolve as a belt about the earth with a decreasing velocity as it spread toward the poles and overcanopied the earth.

7. The smoke or unconsumed carbon that arose from the earth, darkened the upper vapors and formed bands or belts.

8. The moon retarded the rings, causing them to fall upon the earth, and it then receded from our planet.

9. The Archaean metalliferous deposits are so located as to be inexplicable by the old theory of aqueous denudation.

10. The Silurian beds, and particularly the order of their occurrence utterly refutes the idea that they were derived from pre-existing beds.

DEMONSTRATED BY HISTORIC TESTIMONY.

In Gen. 1:7 God made the firmament and divided the waters which were under the firmament from the waters which were above the firmament. According to the Hebrew the atmosphere became an expanse between two bodies of waters, and of course the upper stratum had to move round the earth. In Gen. 1:3,4 light came in and garnished the heavens before the sun was seen.

In the 10th. verse the waters on the earth were called seas, the water above the earth was called the deep, and the Spirit of God moved upon them. "And God said, Let there be light," and light came upon the deep.

In Gen 1:14-19 the sun which existed for ages did not appear in the heavens until after the sun brought forth grass, etc. Then it is plain that some intercepting canopy cut off the direct rays of the sun.

The writer of Genesis did not say the sun and moon shone upon the earth, but he does say the stars did this. According to the Vailan theory this is true, but they shone in from polar regions.

The earth's surface was not heated by the sun's direct rays, but under the overcanopying vapors it must have been warmed, and its temperature equalized by transmitted and diffused solar heat.

CONCLUSIONS.

energy transferred. In the course of 1,000 years, 1,000 square miles of oceanic bottom would be covered to the depth of 240 feet.

This enormous pressure on the underlying rocks is so much transferred energy converted into mechanical heat. This must expand the rocks thus under increased pressure. If this sediment were not borne into the ocean along the Atlantic coast and spread out over vast areas it would be lined with mountains and volcanoes, as that of the Mediterranean sea; but being spread out over an extensive floor it prevents their formation by lateral pressure.

Volcanoes are located where sediments can accumulate, and are doubtless the result of this accumulation. Sixty-five thousand feet of steel blocks piled one upon another would cause sufficient heat to melt the lower ones or reduce them to a plastic state. The lava that issues from a volcano is the deep bed-rock fused by pressure produced by lateral expansion. Accumulating sediments cause rock expansion in some regions, and being removed from others, causes contraction. Expansion elevates the earth's crust; contraction lowers it.

A downfall of water that would raise the ocean fifty feet above its present level would cause an expansion that no rocks could resist, and its lateral pressure must result in mountain making. The New England coast has been elevated in comparatively recent times. The St. Lawrence is so new that it has not yet swept its channel clean.

From Nova Scotia to Florida and around the whole boundary of the Gulf of Mexico are the submerged shore-lines of a former continent. Many miles out the lead-line suddenly plunges from about 100 fathoms to from 200 to 1,500 fathoms. So around the British Isles, the coast of Norway, and that of Northern Europe and Asia. South America, Africa and the Pacific present the same characteristics. The course of a submerged continent has been traced in mid-ocean.

SUMMARY.

The Vailan Theory is proved,

1. By mathematical reasoning and philosophic necessity.

2. By the mineral character and philosophical deposition of strata.

3. By analagous facts relating to other worlds, belted and ringed under the reign of law.

4. By the action of the moon.

5. By the records of man whose ancient writings declare, and re-declare, again and again, the truth of this claim. The first eight chapters of Genesis alone afford proof sufficient if all else failed.

6. The waters on the earth themselves declare the fact.

GEOGRAPHICAL FEATURES.

The first and most important element of the earth's crust is carbon. Of the more than 60,000 feet of aqueous beds there are probably none that it does not enter into as an important factor. It was first driven from the earth by intense heat. The burning world was a smoking world. The unconsumed carbon commingled with the Annular vapors in the form of black, sooty, pitchy matter. This was deposited at the time of the deluge, and the waters that stood in seas, lakes and ponds deposited it as a layer of black, carbonaceous mud upon their bottoms. It may be found in ten thousand lakes planted in the Drift deposits in North America and Northern Europe.

A black carbonaceous soil covers many Western States which were once covered by a vast inland sea. This sea was bounded on the west by the Rocky Mountains; south by the Ozark Mountains and the mountains of Tennessee and Kentucky, and emptied its waters into Lake Michigan.

This great inland sea finally became a fresh-water body. The remains of the mastodon, mammoth and other pachyderms of interdiluvian times, as well as fresh-water shells are found. It made for itself two great outlets, the Mississippi and the St. Lawrence rivers. This inland sea must have been elevated 700 or 800 feet above the ocean, and was surrounded on all sides by walls, and covered an area of at least 500,000 square miles. We must conclude that some great down-rush of waters caused it to break its bounds in two directions at the same time.

The fall of waters supplied the black, sooty carbon that settled to the bottom of the sea, remaining but a few inches thick on the hills, perhaps, but several feet in the valleys, and is the source of the peat bogs.

GLACIAL EPOCHS.

Previous to the glacial record there had closed a long period of perpetual spring. The primitive elephant, and many of his congeners and contemporaries, fed in luxurious forests and grassy plains toward the north pole, which are now covered with glaciers grinding their bones to dust. Northern regions which for untold ages had been covered with tropical vegetation, and animals of innumerable forms, began to be invaded by glaciers which slowly made their way toward the equator.

The only way glaciers are now formed is by vapors wafted over them from adjacent lands warmed by solar heat; but they were not formed that way during the glacial epochs, but by the declension of annular vapors. Glacial ice cannot accumulate extensively now. *It flows*, and cannot be heaped up largely, its rate of motion being proportionate to the slope of its bed. The

We can scarcely conceive of matter anywhere without associating it with living forms. The outermost vapors of the annular system, which fell in the time of Noah, remained on high for unknown millions of years, receiving constant additions of meteoric and cosmic dust from without. As the gaseous envelope that now surrounds our earth contains living organisms, we must believe the annular matter did also, and to a much greater degree.

If Jupiter's belted system had long ago descended to the body of that planet, so that we could gaze upon the continents and seas as we do those of Mars, we would conclude that they swarmed with life. An incomplete world must contain incomplete or primordial life-forms; forms that in time must develop. In yellow snow, dust showers, "blood rains," etc. we have evidence that organic forms are natural accompaniments of the nebulous and elementary forms of matter.

Spider showers are well authenticated. Sometimes the air is filled with their gossamer threads upon which they mount to unknown depths of space, where they live. If spiders can live in the air, descend to the earth and live there for a time, and toads can live for untold ages immured in solid rock, they could live in belts of aqueous and mineral matter. The manner in which organisms have succeeded each other on the earth as revealed by the geologic records demands that the annular system was the cradle of infant life, the propagating beds in which the life-germs were placed by the great Gardener of Nature.

It is as reasonable to suppose that germs took form in water under the creative hand before they fell to the earth as afterward, and when we see that each downfall brought new life-forms which exhibit no specific or generic relation to previous forms, we are forced to admit that either the seed beds of the Annular system provided the undeveloped organisms, or there was a special creation at each period.

In the Silurian age there was an ocean containing heavy calcarious matter; in the Devonian silicious and silicio-calcarious matter; in the Carboniferous carbonaceous matter, and each ocean had its characteristic life-forms. But if all the waters fell at one time, how is it possible for each age to have had an ocean containing characteristic minerals? These characteristic minerals fell with each ring, which marked the ages of geology, destroying previous life-forms and introducing new ones. Eozoic rocks were laid down 40,000 feet thick. Upon these were piled Silurian 65,000 feet thick; on these Devonian rocks 15,000 feet, and then comes 17,000 feet of Carboniferous rocks, each age having characteristic fossils and mineral deposits. As these deposits were laid down by the sea, why do they so widely differ in their composition if they all fell at the same time from above! The Potsdam sandstone underlies the Silurian rocks. It spread from the Canadas to Texas, from the Alleghanies to the Rocky mountains, and probably forms a casement around the globe. It is 8,000 feet thick, and shows a mechanical and rapid accumulation, pointing unmistakably to the downfall of a silicious ring.

The Annular theory admits of the universal eroding power of rivers and waves; the transporting power of currents and strata building from detrital matter. But waves can do nothing unless supplied with matter. Where did they get the crystalline, granulated and infusorial matter to spread over the floor of the Silurian ocean? Great beds of metals have been laid down as regularity stratified deposits which could not have been borne from Archaean terranes.

CARBON STRATA DEPOSITED AS AN AQUEOUS SEDIMENT.

Carbon composing a peat bed is simply unconsumed carbon. The carbon or smoke that arises from every chimney and furnace when measurably shut up from immediate union with oxygen, remains an unburnt fuel precisely the same in kind as the unburnt carbon fuel of the peat bogs. Were we to collect the unburnt carbon from our chimneys in piles, where moisture and air could have free access, it would take fire spontaneously and burn, just as peat dug from the bog sometimes takes fire and burns.

The millions of fires from foundries, volcanoes, etc., are forming fuel wherever soot is formed, and were it not for the ever active oxygen of the air, it would all descend upon the earth as fuel and become incorporated in forming sedimentary beds. This is our claim for the coal, which as unconsumed carbon arose beyond the reach of destroying oxygen, from the heated, glowing furnace of our globe, and in time returned to the earth.

When the plant dies and begins to decay one of its constituent elements, carbon, oxidizes by slow combustion and returns to the air as an invisible gas. It is but accidental when a particle fails to become oxidized and remains as unconsumed carbon. An exceedingly small part of vegetation remains unburnt.

Coal veins, which are from one foot to three hundred feet thick, would make a stratum around the earth ten feet thick. Fifty pounds of coal will yield 10,000 gallons of carbonic acid. Then calling eight gallons equal to one cubic foot the astonishing fact comes out that the coal beds actually draw from the atmosphere an ocean of carbonic acid which would have covered the globe to the depth of 12,500 feet, which would have destroyed all animal life. Even three or four per cent. of carbonic acid in our present atmosphere would be fatal to animal life. Hence it is clear that coal cannot be attributed to vegetable origin.

CONCLUSIONS REACHED.

The following conclusions are clearly deducible:

1. The Annular system was a region of microscopic life and infusorial forms. Coal being deposited by sea-water carried down with it marine forms, and others settled upon its surface.

2. The carbon deposits must have borne down a vast amount of marine vegetation and buried it upon the sea bottom. In swamp marshes the vegetation would have been entirely different.

3. When a carbon fall was borne to the seas and settled where limestone strata prevailed it would indicate great distance from the shore, and here the roof shales of the coal must be necessarily free from land fossils. Coal beds amongst sandstone strata indicate depositions near shore, and may contain land fossils.

4. The coal beds must be more heavily developed toward polar regions, and most free from impurities.

5. All carbon downfalls must have been attended by great cataclysms of snow, or water, or both.

6. A coal vein deposited near a volcano, or mechanical heat arising therefrom would be metamorphosed into heavier and harder forms of carbon. But as all grades must have existed in the Annular system as primitive distillates, all of these forms may be found in lands where no strata disturbance has taken place.

7. The heavy carbon, as the anthracite and semi-bituminous particles would be borne to the deep seas, while the lighter would float into shallow water. Hence a submarine valley might have a deposit of anthracite while a neighboring bed on an elevation might be bituminous.

8. In both northern and southern hemispheres the coal must be more valuable as we proceed from the equator.

9. There must have been carbon falls in all ages, and the first were the purest and the best, while the last to descend must have been the lightest and poorest, and must be found near the surface, or are the foundations of recent peat bogs.

Peat vegetation, or moss known by the generic name of *Sphagnous*, has led many to believe it to be the origin of that product. But these *sphagnous* mosses could never have planted themselves over the medial and colder latitudes if the carbon beds necessary to sustain them had not previously been planted there. If coal and peat are vegetable products they should exist in greater abundance in tropical regions; but they are found in limited quantity there.

IS COAL A VEGETABLE PRODUCT?

The usually accepted theory concerning the origin of coal is that it was formed from an ancient vegetation that grew largely in peat and swamp marshes. This theory the Vailan system overthrows.

Every atom of the great mass of carbon now forming the coal deposits must have been a distilled product of a primitive igneous process before the plant could possibly appropriate it. Every intelligent chemist knows that the great telluric gas furnace of primitive times was competent to produce all the carbon now found in the crust of the earth. Soot, that sometimes takes fire in our chimneys, is deposited in infinitesimal smoke particles. Hence, smoke from burning carbon is simply a fuel which makes it evident that the smoke which arose from the igneous earth was a fuel hydro-carbon. The dark belts of Saturn and Jupiter are doubtless strata of carbon revolving about those planets.

If the Vailan theory is true the graphites and heavier forms of carbon were the first to fall upon the earth after the igneous period was passed, and will be found in its first aqueous beds, and generally unassociated with fossil vegetation. This is precisely what we do find. Both Dana and Dawson bear testimony to the fact that graphite is a very common mineral in the older beds, and that the primitive carbon beds are equal in gravity to that of similar areas in the carboniferous system.

Why no fossil plants in the earlier coal deposits? Because no plants grew at that time. Then we must look for its origin elsewhere than in plants. If coal be a vegetable product, so is graphite. To say that animal organism aided in the process simply adds to the difficulty, since it is carbon that makes the organism and not the organism the carbon. But suppose fossil plants were found in graphite, would it be any more evidence that they formed it than that they formed clay or sandrock in which they are found? The simple fact that organic fossils are found in carbon beds changed to carbon affords no evidence that these organisms made the beds.

We find vegetable remains in coal seams just as we find them in any other rock. A coal plant as a lepidodendron, may begin in the lower clay, and pierce through a coal seam into the overhanging shale and sandstone. In the first it is a clay fossil, in the second a carbonaceous fossil, and in the third a silicious fossil. The fact is the trunk of a tree in an upright position in a coal bed, which is quite common, proves that the coal formed around it rapidly. It would require forty feet of vegetable debris to make five feet of carbon. Some coal seams are 300 feet thick, which would require at least 2,400 feet of vegetable growth in its formation, which is an impossibility. As a vegetable product coal would form very slowly, but from the Vailan system would require but a few hours, or days at most, to lay it down.

Plants found in coal burn with difficulty, which ought not to be true if they contained a resinous sap, or bituminous matter. In many instances you can find a dozen fossil plants in the overlying clay to where you can find one in coal. They are clay fossils because they are imbedded in clay, same as fossils in coal are carbon because imbedded in carbon.

If coal is compressed peat, as some would have us believe, why do we not find fibres running vertically through it? You may examine peat after a pressure of twenty tons to the square inch has been exerted, and yet the vertical structure of the mass will be apparent. Since we find abundance of rootlets running in all directions, vertically as well as horizontally in the under clays of coal beds it is evident that coal is not a metamorphosed peat.

Imagine an expanse of marshes 100,000 square miles in extent, covered with calamites, ferns, sigillaria, lepidodendra remaining motionless for countless centuries, and then suddenly sinking beneath the waves of the sea in order to receive a sea-formed bed for a covering; and in the universal burial to preserve but a few fossils, and they in a horizontal position, while in the clays immediately above and below the coal beds they are found in profusion; that in due time the vast area arose from its baptism, and on the thin layer of clay millions of the same plants grew until they formed another bed of coal, when it sinks again beneath the waves, and this oscillation continued until it had been buried twenty, forty or one hundred times, and you have the old theory of how coal was formed.

But if the old theory concerning the formation of coal is correct, how did it occur that the earth in rising out of the ocean stopped each time in the right place for swamp vegetation to accumulate? According to the highest authority coal is not formed from sea-plants, for they cannot emit any considerable amount of caloric, but it is the product of land plants. Then why do we find coal scattered over a vast area of sea bottom?

The structure of continents show that they have remained such from their first formation. Some of the geologic formations, as the Carboniferous-conglomerates, took place all over the earth at the same time. How could this be except it came from the Annular system?

Were we to have a shower of carbon dust it would settle to the bottom of the sea all over the irregularities of the same. Then sand beds accumulating for ages would settle over it. These would form a greater thickness in some places than in others; hence a succeeding fall of carbon settling upon the ocean floor would not form a bed exactly parallel with the first. This is precisely what we find to be true in the carbon deposits. The distance in coal seams may vary from twenty feet in one place to forty feet in another place in the same neighborhood, which is the result of irregularity in the ocean floor.

Bowlders are found in coal seams which means that coal beds have been formed under water; and if a foreign bowlder that the coal seam was formed at the bottom of the ocean. Bowlders have been found in the middle of coal seams with glacial marks upon them, showing that they have been dropped from icebergs into the forming coal beds at the bottom of the sea. Foreign water-worn bowlders are frequently found in coal beds.

Stratas of coal may be separated by layers of clay not more than half an inch in thickness; how could vegetation take root in so thin a layer of clay sufficient to form the overlying coal seam of probably several feet? Suppose a great carbon fund should float from the Arctic ocean into Hudson Bay. It would settle upon an undulating bottom, and if a flood of muddy water from the surrounding rivers should empty into the bay while the carbon bed was forming, a thin clay bed would be the result. This might continue as long as the carbon was brought from the Arctic regions.

The floating mass of primitive carbon clouds after they entered the atmosphere and floated away for centuries, perhaps, toward the polar regions in their efforts to reach the earth, became a tissue of evolving vegetable organisms and vegetable forms. Take fresh soot from a furnace soon as it is formed, subject it to hot vapors from boiling waters and store it away in an open vessel of water, and you will soon see vegetable and animal organisms start into being. Then why not find organisms in revolving soot clouds in the Annular system?

Marine vegetation exists on the sea bottom, and a carbon sediment rapidly accumulating would certainly involve it.

Under almost all the carbon veins lies a deposit of fire clay. Strange that adjoining a highly combustible bed, a substance should be invariably planted that is so refractory as to be used for crucibles in fusing almost every known metal! In this bed lies involved a profuse marine vegetation, and the preservation of its delicate lineaments proves that it was suddenly involved. It is more generally present under coal veins that are more distant from the tropics, and *invariably* in the most distant ones. The fire clay-dust sublimed in the great telluric crucible arose to commingle with primitive vapors and returned with them. When a carbon fall occurred the clay matter being of greater specific gravity was the first to find its way to the ocean floor.

This fire clay is found under beds of primitive graphite where no vegetation is involved, and therefore cannot be a vegetable distillation. It is found where glacial action is unknown, and cannot be mud pulverized by moving ice. Every one of the more than seventy coal seams of the Nova Scotia regions has its characteristic clay-bed. When we see trees standing in and surrounded by this clay we are forced to admit a rapid accumulation.

Limestone is a deep sea formation and the Vailan system demands that standing trees should not be found in it. Only such limestone formation or strata as were deposited as mechanical precipitation could be formed in shallow waters, especially in regions beyond the tropics. A limestone stratum deposited among shore deposits or continental detritus points directly to Annular origin and vegetable fossils will occur in the upper clays. Here geologists have an opportunity to prove or disprove the Annular problem.

Coal and peat are not found in the tropics where they ought to be found if vegetation produced them. And if they could be found there it would sweep the Vailan system from its foundations. They are found, however, just where this system says they must be found. Why is peat found in the ocean, and in the thousands of lakes and ponds where no peat vegetation is now growing? Suppose we find a peat bed forty feet thick, it must have been at one time a lake with forty feet of water, and how did the peat begin to grow? Peat forms slowly and the rains and storms would have worked mud, etc., more rapidly into it than the peat would have filled it. It would neither have grown from the top nor from the bottom. The foundation carbon fell from the Annular fund.

METAMORPHISM OF CARBON BEDS.

When bituminous or lignitic coal, or even peat is subjected to a sufficient degree of heat it is converted into hard coal and sometimes into graphite. From this source some conclude that anthracite and all hard coals are metamorphosed beds of soft carbon. But how about the vast beds in aqueous crusts hundreds of miles from any igneous agencies? All anthracite coal changed from bituminous coal will contain a greater per cent. of ash than the coal from which it is derived. If it does not it is evidence that it never was bituminous coal.

Let us suppose a heavy fall of Annular carbon in the north Atlantic ocean, and that the Appalachian mountains were again under the sea. The carbon carried by the ocean currents southward would fall to the sea bottom in the more quiet waters. The heavy or anthracite dust would reach the bottom in deep waters where the lighter forms would not. Before the Appalachian upheaval, the eastern base of the system was farther out in the sea, and was in deeper waters than the western. The constitution of the coal itself, the condition of the sea bottom (sloping from the coast to the deep sea) point harmoniously to the annular origin of the carbon beds. The bituminous dust not being able to directly settle with the anthracite remained longer in suspension which accounts for its greater amount of ash. The farther south it floated, the more impure it became. The heaviest beds of anthracite will be found in the northern part of the great plateau, and principally in British America if the Vailan theory is true.

Fossil plants in coal are generally mineralized charcoal, and are difficult of combination. If the bed was composed of vegetable production the same difficulty would certainly characterize the mass. Hence the plant is simply a foreign body in a bed of mineral carbon. Coal seams have become so hard as to be planed off by eroding forces directly after being laid down, or before heavy beds had accumulated over them. Thus they could not have been formed by vegetable peat.

TERTIARY COALS.

Extensive coal beds in Asia are probably Tertiary, while the vast carbon beds among the Rocky Mountains, and underlying the vast plain to the west of these mountains, were formed in the Tertiary period. The Rocky Mountain plateau on which the coal beds are planted existed as a sea bottom over which the waters from the Arctic world rolled during the Tertiary period. The Rocky Mountain region was then sleeping in the sea.

The Tertiary beds reach from Mexico to the Arctic ocean, proving that currents ran toward the equator along the valley of the McKenzie, bearing into southern waters whatever fell from the upper world. It is thus easy to see how the vast expanse of this western world became the receptacle of Tertiary carbon. Finding no Tertiary coals on the Eastern border of our continent we are led to believe that a narrow continent stretched from America to Europe across the present bed of the Atlantic and hindered the flow of carbon along the Atlantic seaboard. It is now conceded by geologists that such an isthmus of land reached from Newfoundland to the shores of Europe during the Tertiary period. This being true a vast fund of carbon must lie at the bottom of the North Atlantic.

If these later coals had been formed out of vegetation growing in great continental swamps, the same opportunity was afforded by the southern sea borders for this swamp vegetation. And so from Long Island to the Rio Grande. Why then do we not find it if coal is of vegetable origin? If the vast fund of the lignitic coals is a vegetable production it was present in the Tertiary atmosphere as a deadly poison. But at that time both land and sea were full of air-breathing mammals and monsters showing conclusively that it was not there in such a condition.

DEDUCTIONS.

1. The plant when subjected to a proper mode of distillation is made to yield carbon in various allotropic forms. So of any mineral that has carbon in its constitution. These forms of carbon were placed in the crust of the earth after the primitive fires had died out.

2. All such primitive distillations existed in the atmosphere of the incandescent earth.

3. This matter as it declined and mingled with the atmosphere in after ages, changed from the ring to the belt form, and overcanopied the earth and fell largely in regions outside the tropics.

4. The heavier forms of carbon fell largely in the earlier ages; though all sections of the system must have had some of each form.

5. All ages were more or less characterized by carbon falls, and no age could be exclusively carboniferous.

6. Carbon falling directly into the ocean would separate into heavier and lighter forms and settle accordingly in higher or lower elevations of sea bottom, thus explaining why different forms of coal are found in the same proximate horizon.

7. The earliest or heavier forms are free from organic remains, and must therefore be a primitive distillation. The other carbon beds by their associated strata; by their involved vegetation and other organisms; by accompanying clay-partings; by involved glacial drift; by latitudinal gradation in quantity of ash and specific gravity; by characteristic absence from the tropics and the heavy deposits in higher latitudes; by synchronous formation in all continents; by their evident formation in the very lap and bosom of the glacier and in ice and flood; by the fact that they are bituminous, oily hydro-carbons, and by a multitude of inconsistencies and impossibilities involved in the vegetation theory, have been shown to be actual sedimentary deposits, and therefore a primitive product.

Since then there is not a feature connected with the formation of coal that is not readily explained by the primitive carbon theory; not one that philosophic law does not resolve into harmony with Annular declension without even the show of conflict; and since vegetarians are forever stumbling upon inexplicable difficulties—bowlders, pebbles, undulations, slopes, ripple-marks, clay-partings, cannel-coal inseparably joined with bituminous coal, anthracites with less amount of ash, marine impurities, carbon planted in Archaean beds, air-breathing animals among Tertiary coals, carbon dredged from the ocean, dug from the frozen world, and innumerable other objections over which they can not climb, the vegetation theory can not be true.

ANNULAR DOWNFALL IN THE TERTIARY OCEAN OF THE NORTHERN HEMISPHERE.

If the Vailian theory claims are valid the beds in the Rocky Mountain Tertiary must present the following features: The Cretaceous period having been brought to a close by a down-rush of waters and snows in the northern hemisphere, a stream of water pouring southward must to a great extent have been a fresh-water current, and those deposits in the extreme northern beds of the Rocky Mountain region must be largely fresh-water accumulations. Those in the middle of this region must be to a less extent fresh-water; perhaps sometimes fresh and again marine, owing to changes in currents, etc., and the two be commingled, while in the southern part the beds must be almost exclusively marine. Fortunately for the Vailian theory these demands are fully met. The waters of this vast region communicated with the Arctic ocean, probably by way of the present depression in British America, along the valley of the McKenzie river, while south it communicated with the Gulf of Mexico.

Here was a sea forty times larger than Lake Erie. Where did the water come from that made the northern part fresh, the middle part brackish and the southern portion marine? The Tertiary of the Pacific Coast is marine; so is a larger portion of the Atlantic border. Doubtless Davis Strait poured a volume of fresh-water from the polar world into the Atlantic, for there is the same commingling of marine and fresh-water shells on the northeast coast, while in the northern part they are exclusively fresh-water species. Rivers could not have done this, for all the rivers from Delaware Bay around the coast of the Gulf of Mexico were not sufficient to lay down fresh-water Tertiary. Admit that the vast polar ocean of the Tertiary period was a body of fresh-water, and all difficulties disappear.

Geologists admit that in the Tertiary period mountains were made on every continent, that there was a world-wide disturbance of strata, and the most complete extermination of species on record. The Cretaceous world was swept by a mighty cataclysmic wave, and its animals were buried in the detrital mass swept from the land into the seas and formed the lower Eocene beds. Nothing of which we can conceive could do this but a downpour of Annular waters. One-third of North America, a great part of Northern Europe, nearly all of Siberia, much of China, and other parts of Asia were apparently synchronously submerged beneath *fresh-water.*

The ocean of fresh-water proves the augmentation of snows from the great super-aerial fund. The Cretaceous age closed by excessive and unusual refrigeration. The transported blocks of stone found in the Upper Cretaceous and Lower Tertiary point to a northern origin. The evidence is overwhelmingly in favor of an Annular fall of waters in the north polar world at that time.

Existing continents were submerged under Cretaceous waters. The Rocky Mountains, Andes, Alps and Himalayas were either unborn or in their infant stage. But some mighty barrier was raised that rolled the Cretaceous waters southward, and made an isolated fresh-water ocean on the north. It was the great Atlantic plateau reaching from Newfoundland to Ireland, which is known by actual soundings and other evidence to be a submerged table land.

It was raised from the deep at this very time and stood for uncounted milleniums as dry land.

Suppose an ice cap 5000 feet thick should suddenly cover the Arctic world. It would press that part of the globe inward and downward upon itself even if the planet were solid to the centre. It would render the rocks plastic and they would be pushed under the continents causing the crust of the earth to rise into mountains in many places. Just what occurred in Cretaceous and Tertiary times.

We can trace the shore-line of an almost limitless fresh-water sea around the whole hemisphere in Tertiary times, showing that the Arctic ocean was a wide expanse of fresh-waters. This leads to the positive and permanent establishment of the Vailian or Annular theory.

APPENDIX.

THE LAST ADVANCE OF GLACIERS.

The last downfall of exterior vapors was at the time of Noah, and produced the deluge. These vapors naturally gravitated toward the polar regions and falling there as snows would accumulate as glaciers, their magnitude and extent corresponding with the amount of falling snows. It is evident if there ever was an Eden climate upon the earth its destruction was brought about by a change of climate. If the Deluge was a collapse of the last remnant of upper waters the latter must have begun to fall in polar regions many centuries previous.

The Eden world suffered a change of climate during the Adamic age, for the race that dwelt naked in Eden became clothed in the skins of animals. If this infant race dwelt naked the climate was warm. If afterward it became necessary to be clothed with the skins of animals it certainly had become cold. If the cold increased it was probably caused by the fall of snow in polar regions. The physical condition of the antedeluvians and their environment depended on the conditions of the upper vapors. Hence, polar glaciers began to advance in Edenic times.

Glaciers advanced slowly, and are still advancing. Eight hundred years ago Greenland was not the frigid land it now is. The Icelanders and the Northmen sailed through northern seas in the interest of commerce where now our hardiest seamen with iron-clad vessels scarcely dare to venture. They pushed forward commercial enterprises into lands that are now inhospitable and uninhabited.

The present glaciation of polar worlds is but the result of the last declension of outward vapors. The great ice caps of polar regions are moving toward the equator and are constantly diminishing. It is possible that we are approaching a day when the last ice berg will be borne toward the tropics, and the last glacier will melt, and a more genial climate pervade the greater portion of the earth.

LONGEVITY OF THE ANCIENTS.

According to the biblical account people lived to be 800 and 900 years old. This was principally because of the modification of solar energy. Man's physical environments impelled long life; and his longevity diminished immediately after the upper deep fell and the sun began to pour his beams upon the race; his environment evidently changed with that event. In a few generations after the flood man died at the age of 120 or 100 years, and finally at three score and ten.

LETTER FROM PROF. I. N. VAIL.

MY DEAR DR. BOWERS: I have read with much interest thy compendium of "The Earth's Annular System," as published by me in 1886. A synopsis of that work can give but a meager idea of the grand conception of the annular evolution of the earth. "The Annular Theory" stands on the immutable truth that worlds evolve according to invariable law.

This compels us to admit that all worlds are made alike, in the general changes they undergo. Just as a bud evolves into a flower of the most delicate construction and architectural order, so a world launched from the same designing Hand must move in the same line of eternal order, and under the law of natural uniformity develop and grow into a completed world.

This also leads us to the conclusion that if one world possess at any time an annular system, then all worlds must possess a similar appendage during some period of their existence. Consequently that simple fact that the planet Saturn possesses at this time an annular or ring system is proof that the earth once had a similar appendage. For we must either admit this truth or we must admit that the planet Saturn has not evolved thus far along a line of nature's uniformity, but is today a victim of accidental conditions. This law refuses to admit.

But "The Annular Theory" does not rest on these grounds alone. A universe of *invariable order* pronounces it an immutable truth. The judgment of the chemist and philosopher is positive that a rotating world cannot pass from the molten state to the present condition of the earth without undergoing annular changes.

Since the publication of "The Earth's Annular System" I have had opportunities of examining more minutely the subjects treated of therein and have secured the most overwhelming evidence that the theory there proposed is in the main correct and will stand the test of all time. I have found, outside the realm of physical science, the most positive evidence that primitive man actually saw at least two rings revolving about the earth, named them and worshiped them as gods. These relics I have rescued from the wreck of ages, and *with* these I will prove the fact that this earth once had a complex system of Saturn-like rings.

Thus in the end the geologist and astronomer will be compelled to admit its truthfulness whether they desire to or not. I have found among the ruins of ancient Egypt, Babylonia, India and China annular fossils, the identification of which settles at once and forever this great question.

Again, I need not point the geologist to the mysteries of the glacial epochs, which grow darker and darker as he looks for a competent cause for their production. He must know that the great ocean of vapors that hovered for unknown time over the earth in the loftiest heights of the atmosphere, such as now are seen on two of our neighbor planets, could not have fallen to the earth without covering it in the higher latitudes with measureless masses of snow, resulting in excessive refrigeration. I need but point him to the fact, proven by the coast surveys of the world, that the oceans have encroached upon the land to such an extent since the last glacial epoch that they stand now fully thirty fathoms deeper than they did in pre-glacial times. I need only point him to that grand clock-work of worlds shining from the firmament— every scintillating point, every rolling sun, is a witness of nature's eternal order, and proclaims that uniformitarian principle of world evolution, by which the philosophic investigator must stand. The geologist must build on this rock of *uniformity* in the evolution of worlds. The earth has evolved along *this* line, and the wreck of annular conditions is seen on every page of its rocky volume.

In the year 1875 I published a little volume entitled "The Earth's Aqueous Ring." In it I stated my convictions, and gave reasons therefor, that all the glacial periods the world ever saw were produced by supra-aerial vapors descending from an annular system that revolved about the earth from the remotest geologic ages to the flood of Noah, which was itself produced by the fall of the last remnants of those upper waters. These claims I am fully prepared to substantiate, whatever opposition may be brought against them.

ISAAC N. VAIL,

ELSINORE, Cal., July 6, 1892.

Milton Keynes UK
Ingram Content Group UK Ltd.
UKHW030744071024
449371UK00006B/570

9 789362 099518